WITHDRAWN

JEFFERSON COUNTY PUBLIC LIBRARY

The Moon Has Phases

Copyright © 2018 by Belinda M. Sanders

All rights reserved. No part of this book may be reproduced or transmitted in any form or by any means without written permission from the publisher and author.

Additional copies may be ordered from the publisher for educational, business, promotional or premium use. For information, contact ALIVE Book Publishing at: alivebookpublishing.com, or call (925) 837-7303.

Book Design by Alex Johnson

ISBN 13
978-1-63132-061-3
ISBN 10
1-63132-061-0

Library of Congress Control Number: 2018965824
Library of Congress Cataloging-in-Publication Data is available upon request.

First Edition

Published in the United States of America by ALIVE Book Publishing and ALIVE Publishing Group, imprints of Advanced Publishing LLC
3200 A Danville Blvd., Suite 204, Alamo, California 94507
alivebookpublishing.com

Printed in the United States of America
10 9 8 7 6 5 4 3 2 1

The Moon Has Phases

Belinda M. Sanders

ABOOKS
Alive Book Publishing

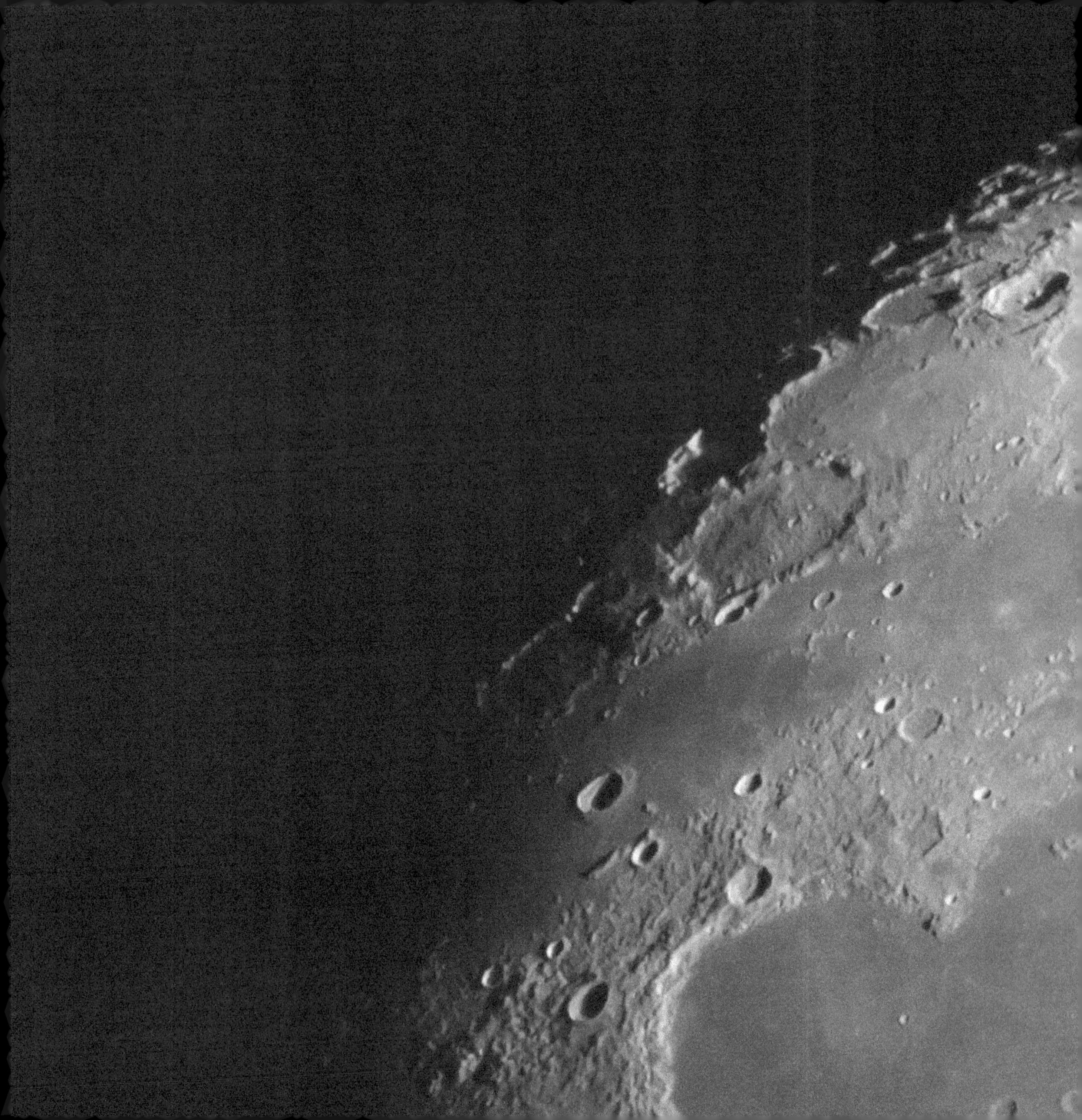

I dedicate this book to all the moon babies I have taught.

There

High in the sky

Earth is what it orbits

Making its journey slowly

First Quarter
Waxing Gibbous
Waxing Cresent
Full Moon
Orbiting the earth
New Moon
Waning Gibbous
Waning Cresent
Last Quarter

Oxygen isn't available on the moon

Not much gravity either...

Half of it, is visible to the eye

Always appearing to reflect light

Shining from the sun

Pulling night across the sky

Hovering at the horizon's edge

An arc, or

Sliver-like sickle

Eclipses the

Sun

The end!

Quick Check Moon Facts

- What does it mean to say that the moon orbits earth?
- If the moon is crescent-like in shape, what phase of the moon are we seeing?
- What causes the moon to reflect light?
- What causes the moon to eclipse the sun?
- Why is the moon devoid of oxygen?
- Why is there less gravity on the moon?
- Why is a New Moon nearly "invisible" in the night sky?

Moon Facts to Research

- How many phases does the moon have?
- When does the New Moon appear?
- How far away from Earth is the moon?
- How long does it take for the moon to orbit Earth?
- Why does the moon have no oxygen?
- What is a lunar eclipse?
- Name the first astronaut to walk on the moon.
- Who is Dr. Ronald McNair?

Words & Concepts to Know

- Arc
- Crescent moon
- Earth
- Eclipse
- First quarter moon
- Full moon
- Gibbous moon
- Gravity
- Horizon
- Hovering
- Last quarter moon
- Lunar
- Lunar eclipse
- Moon
- New moon
- Orbit
- Oxygen
- Quarter Moon
- Reflecting
- Sun

Blue Moon

_____

_____

_____

_____

_____

_____

_____

Research interesting facts about this type of moon.
List your findings on the above lines.

About the Author

Belinda M. Sanders is an elementary school educator teaching in Oakland, California. She enjoys becoming one with the books she reads. She considers herself a gypsy child of grace. She lives in Northern California.

ABOOKS

ALIVE Book Publishing and ALIVE Publishing Group
are imprints of Advanced Publishing LLC,
3200 A Danville Blvd., Suite 204, Alamo, California 94507

Telephone: 925.837.7303
alivebookpublishing.com

CPSIA information can be obtained
at www.ICGtesting.com
Printed in the USA
LVHW072047151221
706292LV00006B/468

1-22

JEFFERSON COUNTY PUBLIC LIBRARY